Red-Tailed HAWKS at Big Bend

written by
FRANCES E. RUFFIN

Pearson Education, Inc. 330 Hudson Street, New York, NY 10013

© 2020 by Pearson Education, Inc. or its affiliates. All Rights Reserved. Printed in Mexico. This publication is protected by copyright, and permission should be obtained from the publisher prior to any prohibited reproduction, storage in a retrieval system, or transmission in any form or by any means, electronic, mechanical, photocopying, recording, or otherwise. For information regarding permissions, request forms and the appropriate contacts within the Pearson Education Global Rights & Permissions Department, please visit www.pearsoned.com/permissions/.

Photo locators denoted as follows Top (T), Center (C), Bottom (B), Left (L), Right (R), Background (Bkgd)

Cover Robbie George/Getty Images; Back Cover Danita Delimont/Getty Images.

1 Arvo Poolar/Getty Images; 3 (Bkgrd) Photography by Deb Snelson/Getty Images, (C) Arvo Poolar/Getty Images; 4 Walter Bibikow/Getty Images; 5 (B) Danita Delimont/Getty Images, (T) Carol Polich Photo Workshops/Getty Images; 6 Inge Johnsson/Alamy Stock photo; 7 Ondrej Prosicky/Age Fotostock; 8 Genevieve Vallee/Alamy Stock Photo; 9 John Cancalosi/Alamy Stock Photo; 10 Johann Schumacher/Alamy Stock Photo; 12 Robert Shantz/Alamy Stock Photo; 13 Robbie George/Getty Images; 14 Robbie George/Getty Images; 15 Johann Schumacher/Alamy Stock Photo; 16 John Cancalosi/Age Fotostock; 17 Barbara Friedman/Getty Images; 18 Danita Delimont/Alamy Stock Photo; 19 Danita Delimont/Alamy Stock Photo; 20 Danita Delimont/Getty Images; 21 Danita Delimont/Getty Images; 22 Deb Snelson/Getty Images; 23 Melinda Fawver/Shutterstock; 24 Rolf Nussbaumer Photography/Alamy Stock Photo.

PEARSON and ALWAYS LEARNING are exclusive trademarks owned by Pearson Education, Inc. or its affiliates in the U.S. and/or other countries.

Unless otherwise indicated herein, any third party trademarks that may appear in this work are the property of their respective owners and any references to third party trademarks, logos or other trade dress are for demonstrative or descriptive purposes only. Such references are not intended to imply any sponsorship, endorsement, authorization, or promotion of Pearson's products by the owners of such marks, or any relationship between the owner and Pearson Education, Inc. or its affiliates, authors, licensees or distributors.

ISBN-13: 978-0-328-94181-0
ISBN-10:   0-328-94181-6

Kee-eeeee-arr!

A hawk flies over the desert.

Far below, desert animals are quiet. They know the hawk is near.

A jackrabbit hides under a bush.
A lizard stops eating red ants.

The desert is in a park called Big Bend. A big river bends and flows here.

The hawk is a red-tailed hawk. Look at it soar! Let's learn more about it.

Red-tailed hawks live in many places. They build big nests.

Some nests are in trees.
Some nests are on rocks.

The mother hawk lays eggs. Both the mother and father keep the eggs safe.

The mother and father sit on the eggs. They take turns.

The eggs hatch. Baby hawks come out. The babies are called chicks.

The chicks are small.
They cannot fly yet.

The mother and father care for the chicks. They bring food to the nest.

The chicks peep. They are hungry. They open their mouths. The mother drops food in.

The chicks get bigger. They grow stronger.

They grow brown feathers. Soon the chicks will fly.

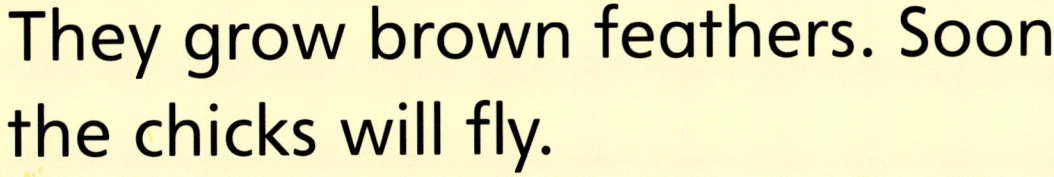

The young hawks are ready. They start trying to fly.

They flap their wings. They are practicing.

A young hawk jumps. Flap! Flap! Flap!

It takes a few tries. Wheeeee!

Soon the young hawks can fly.

Kee-eeeee-arr!

They fly over Big Bend.

They look for food. They learn to hunt. Maybe they will catch a jackrabbit or a lizard.

The hawks become adults. They will have their own families.

Big Bend is their home.